Arctic Tern Migration

By Susan H. Gray

21st Century Junior Library

Cherry Lake Press

Published in the United States of America by Cherry Lake Publishing Group
Ann Arbor, Michigan
www.cherrylakepublishing.com

Reading Adviser: Marla Conn, MS, Ed., Literacy specialist, Read-Ability, Inc.
Content Adviser: Sheila K. Schueller, Ph.D., Lecturer and Academic Program Specialist, University of Michigan School for Environment & Sustainability
Photo Credits: ©Philip Pilosian/Shutterstock.com, cover; ©Dominik Belica/Shutterstock.com, 4; ©Curioso/Shutterstock.com, 6; ©Sadie P Photography/Shutterstock.com, 8; ©photolike/Shutterstock.com, 10; ©Rocky Grimes, 12; ©Himanshu Saraf, 14; ©LouieLea/Shutterstock.com, 16; ©Yuriy Bartenev/Shutterstock.com, 18; ©Tony Brindley/Shutterstock.com, 20

Cherry Lake Press is an imprint of Cherry Lake Publishing Group.

Library of Congress Cataloging-in-Publication Data

Names: Gray, Susan Heinrichs, author.
Title: Arctic tern migration / by Susan H. Gray.
Description: Ann Arbor, Michigan: Cherry Lake Publishing, [2021] | Series: Marvelous migrations | Includes index. | Audience: Grades 2-3
Identifiers: LCCN 2020002732 (print) | LCCN 2020002733 (ebook) | ISBN 9781534168572 (hardcover) | ISBN 9781534170254 (paperback) | ISBN 9781534172098 (pdf) | ISBN 9781534173934 (ebook)
Subjects: LCSH: Arctic tern–Migration–Juvenile literature.
Classification: LCC QL696.C46 G74 2021 (print) | LCC QL696.C46 (ebook) | DDC 598.3/381568–dc23
LC record available at https://lccn.loc.gov/2020002732
LC ebook record available at https://lccn.loc.gov/2020002733

Cherry Lake Publishing Group would like to acknowledge the work of the Partnership for 21st Century Learning, a Network of Battelle for Kids. Please visit http://www.battelleforkids.org/networks/p21 for more information.

Printed in the United States of America
Corporate Graphics

CONTENTS

Terns are seabirds that are known for their forked tails.

The End of Summer

August has arrived. **Arctic** terns are getting restless. Their eggs have hatched. They have raised their chicks. They've been feeding daily on tiny fish and **crustaceans** in the ocean. But things will change soon. The days are getting shorter, and temperatures in the Arctic will begin dropping. It's about time to leave.

The terns will soon begin a long **migration** south. They arrived here last

Females can lay one to three eggs at a time.

May. They spread throughout northern Europe, Russia, and North America. For the last few months, they enjoyed temperatures of about 60 to 75 degrees Fahrenheit (16 to 24 degrees Celsius). Handsome males **courted** picky females. Pairs mated. And females laid **speckled** eggs. Over the next 7 weeks, chicks hatched, grew up, and learned to fly.

Ask Questions!

Arctic terns lay their eggs on the ground. Find some pictures of their eggs. Is there any advantage to having speckled eggs? Would it matter if the eggs were a solid color?

Arctic terns often mate for life.

Heading South

Not all terns take the same migration pathway. Those living near Alaska fly alongside North and South America. Birds near Greenland and Europe head toward Africa's west coast. Some of these terns will split off and fly toward South America. Others will continue closer to Africa. Terns in northeastern Russia will fly east, then south past Australia. No matter the route, all are headed to **Antarctica**.

Arctic terns do not fly in a straight line. They zigzag and curve.

Terns take 3 months or more to make the trip. Their curving pathway from north to south can cover more than 21,000 miles (33,796 kilometers)! They do not fly nonstop. Instead, they have specific stopover points where they must rest and refuel. Terns will stay almost a month at these stops before resuming their travels.

Make a Guess!

Scientists study photos of the ocean taken from space. They see that terns stop where ocean plant life is rich. But terns rarely eat plants. So why would they stop at these places?

Parents bring fish to their chicks until the chicks are old enough to catch their own food.

In November, terns begin arriving in Antarctica. The Weddell Sea is one of their favorite spots. It lies between Antarctica and South America. The sea **teems** with food for the terns.

These birds are very good at fishing. To catch its food, a tern **hovers** just above the ocean's surface. Then, it dives into the water. It swallows the fish whole.

Look!

Fish-eating birds swallow their food headfirst. Look at the body of a fish. Why is it best to swallow a fish this way?

Arctic terns are good gliders. They can even sleep while gliding!

Return Trip

From about November to April, terns spend their time feasting in Antarctica. Here, 24 hours of sunlight provide food for plenty of **plankton**, crustaceans, and fish.

Think!

Migrations require a lot of energy. What are some ways that terns might be able to save energy on their trips?

Arctic terns are protective of their nests. They will swarm and dive on **predators** or people who get too close.

The trip south takes at least 3 months. The trip back to the Arctic, however, takes about half as long.

How do the birds return so quickly? They take advantage of wind currents. In the Southern **Hemisphere**, winds tend to blow in a counterclockwise pattern. In the northern half of the world, winds blow clockwise. By gliding with the wind, the terns save energy and fly faster. In May, they begin reappearing in the Arctic.

Arctic terns are social birds. They travel together in a flock.

To the Moon!

Scientists study these fantastic migrations. They attach tiny tracking devices to the birds' legs. The devices record their movements. This is how they discovered one tern whose journey was truly astonishing. It had flown almost 60,000 miles (96,561 km) in 1 year!

Terns can live more than 30 years. Imagine if a bird flew 50,000 miles (80,467 km) every

Arctic terns experience more daylight than
any other animal in the world.

year for 30 years. It would cover 1.5 million miles (2,414,016 km) in its lifetime. That's more than three trips to the moon and back!

It is incredible what Arctic terns accomplish in 12 months. They travel tens of thousands of miles. They migrate farther than any other animal on Earth. They visit every ocean on the planet. They know just where to stop, rest, and feed. And they return to the same spots year after year. These are truly amazing birds!

GLOSSARY

Antarctica (ant-ARK-tih-kuh) the landmass around the South Pole

Arctic (AHRK-tik) pertaining to the area around the North Pole

courted (KORT-id) behaved a certain way in order to find a mate

crustaceans (kruh-STAY-shuhnz) animals such as crabs and lobsters that have an outer shell and no backbone

hemisphere (HEM-ih-sfeer) half of the globe

hovers (HUHV-urz) remains in one place in the air, without moving in any direction

migration (mye-GRAY-shuhn) movement of animals from one region to another and back again

plankton (PLANGK-tuhn) tiny plants and animals that drift or float in oceans and lakes

predators (PRED-uh-turz) animals that hunt and eat other animals

speckled (SPEK-uhld) covered with small, irregular spots or patches of color

teems (TEEMZ) to have large amounts of something

FIND OUT MORE

BOOKS

Catt, Thessaly. *Migrating with the Arctic Tern*. New York, NY: PowerKids Press, 2011.

Schuetz, Kari. *Arctic Tern Migration*. Hopkins, MN: Bellwether Media, 2018.

Tunby, Benjamin. *The Arctic Tern's Journey*. Minneapolis, MN: Lerner Publications, 2018.

WEBSITES

National Geographic Education—To the Ends of the Earth
https://www.nationalgeographic.org/article/ends-earth/?utm_source=BibblioRCM_Row
Find information on the tern's migration, including how it saves energy by gliding.

YouTube—Arctic Tern Migration
https://www.youtube.com/watch?v=bte7MCSBZvo
Watch an excellent story about one scientist's study of the tern's migration and the problems he encountered.

INDEX

ABOUT THE AUTHOR

Susan H. Gray has a master's degree in zoology. She has written more than 170 reference books for children and especially loves writing about animals. Susan lives in Cabot, Arkansas, with her husband, Michael, and many pets.